HUBBLE

14 FRAMEABLE PRINTS

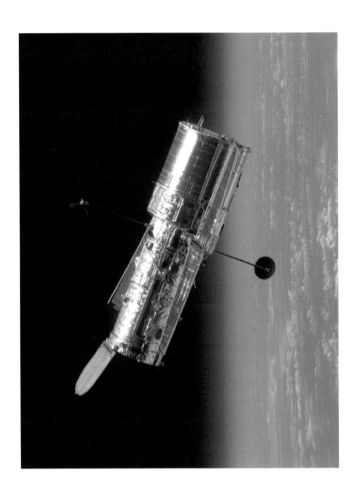

The Hubble Space Telescope is an icon of modern science. It is a technological wonder of our age that has fascinated millions of people with its beautiful images and intriguing discoveries about the Universe around us. It has revolutionized the science of astronomy, revealing hitherto unknown secrets of planets, stars and galaxies that present us with a perspective of the Universe that we could never previously have imagined.

As well as capturing stunning close-up pictures of the planets in our own solar system, Hubble has looked deep into the Milky Way and beyond, taking pictures of rare, massive stars one hundred times larger than our own Sun, and star-forming nebulae that extend across dozens of light years. It has allowed scientists to witness some of the most destructive forces in existence as galaxies collide, and to observe the huge amount of energy released when stars reach the end of their lives. Hubble's Deep Field images are among the most scientifically important images ever taken, showing the most distant and most ancient galaxies in the visible Universe.

Since its launch in 1990, the Hubble Space Telescope has transmitted images in spectacular and unprecedented detail, transforming the way we see the Universe and our place within it. From our nearest interplanetary neighbours to the very edge of the Universe and the dawn of time, Hubble's observations have helped us to answer questions about the age of the Universe, how planets form and even the nature of the supermassive black holes that dwell at the centre of galaxies.

Quercus Editions Ltd
7th Floor, South Block
55 Baker Street
London
W1U 8EW

First published in 2011

Copyright © 2011 Quercus Editions Ltd

ISBN: 978-1-84866-177-6

Printed and bound in China

10 9 8 7 6 5 4 3 2

JUPITER

JUPITER

The first of the giant planets is also the largest world in the solar system, an enormous ball of lightweight gases so big that it could fit every other planet inside it with room to spare. Jupiter's great size makes it easy to view from Earth, and the HST has tracked many features of its constantly changing atmosphere.

Astronomers aren't quite sure whether Jupiter has a solid core – if it does, then it's probably no larger than Earth, and almost irrelevant to what goes on in the layers above. However, huge amounts of heat are generated within the planet, probably as a result of gradual contraction of the deep interior under the influence of gravity. The planet gives out about twice as much energy as it receives from the Sun, driving powerful weather systems in the outer atmosphere.

Colourful layers of cloud are formed where various chemicals condense from gas to droplets at different heights in the atmosphere. The deepest clouds are dark blue and rich in water ice, while mid-level brownish-red clouds contain compounds such as ammonium hydrosulphate, and higher-level white clouds are formed from pure crystallized ammonia.

SHARPEST VIEW

Shortly after Servicing Mission 4 in 2009, Hubble turned the newly fitted Wide-Field Camera 3 towards Jupiter and captured this stunning image. The new dark spot in the southern hemisphere was still rapidly expanding following the impact and breakup of a comet or asteroid in the atmosphere. Photographed across a distance of 370 million miles (600 million km), Hubble was nevertheless able to resolve details as small as 75 miles (120 km) across.

Picture: NASA/ESA/M.Wong (Space Telescope Science Institute, Baltimore, Md.)/H. B. Hammel (Space Science Institute, Boulder, Colo.)/Jupiter Impact Team.

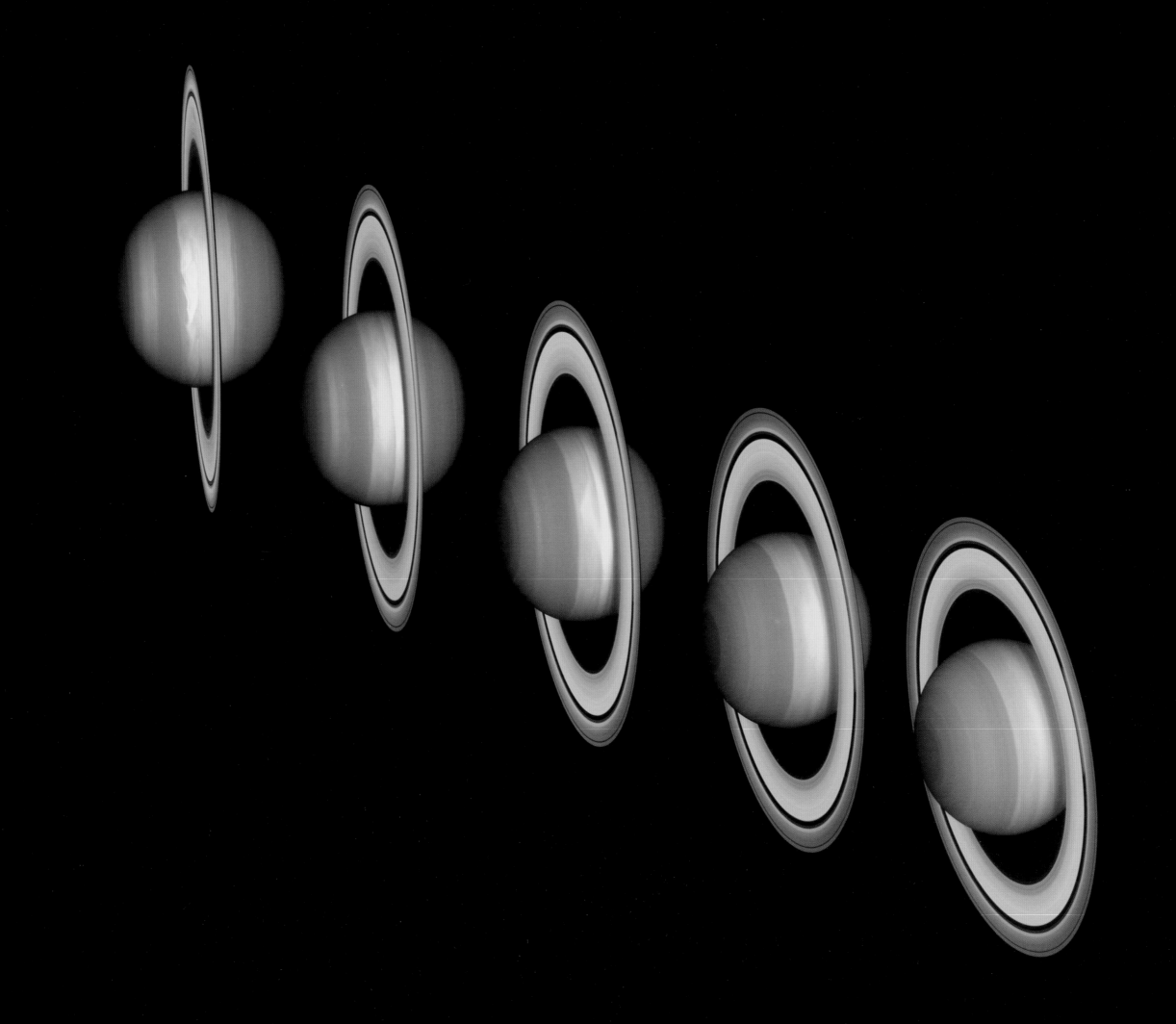

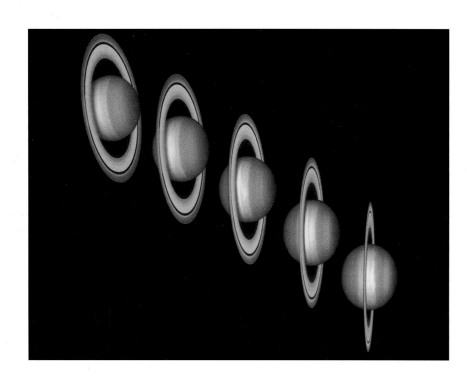

SATURN'S RINGS

Each of the giant planets has a ring system, but Saturn's is by far the most impressive. From Earth, the rings appear to consist of a handful of concentric platters, but the Hubble Space Telescope's superior vision can start to resolve the thousands of individual ringlets that compose each apparently uniform plane.

Astronomers were shocked when the first spaceprobes to Saturn revealed a wealth of unexpected fine detail within the rings. Each turned out to be made from countless narrow ringlets, pushed this way and that by the gravitational influence of 'moonlets' within the rings, and larger moons beyond them. Faint new ring structures are still being discovered even now.

The reason Saturn's main rings appear so spectacular is because of their chemical composition – they are mostly made from highly reflective water ice. Some astronomers believe that such bright rings will inevitably 'tarnish' with age, and so the system must be relatively young. However, most believe that the rings probably formed early in Saturn's own history, and are kept 'fresh' by constant recycling within the rings.

But where did the ring material come from in the first place? It's possible that the main structures are simply surviving debris from Saturn's formation, prevented from coalescing into a moon by Saturn's own gravity. However, it's also possible that they are the pulverized remains of a substantial moon, perhaps over 180 miles (300 km) across, broken up by a major comet impact perhaps 4 billion years ago.

DANCE OF THE RINGS

Saturn is tilted on its axis at almost the same angle as Earth itself, and so we see different aspects of the planet and its rings as it travels around its orbit. This sequence of Hubble images captures the rings as they 'open up' between 1996 and 2000, bringing the southern hemisphere into view.

Picture: NASA/Hubble Heritage Team (STScI/AURA) Acknowledgement: R.G. French (Wellesley College).

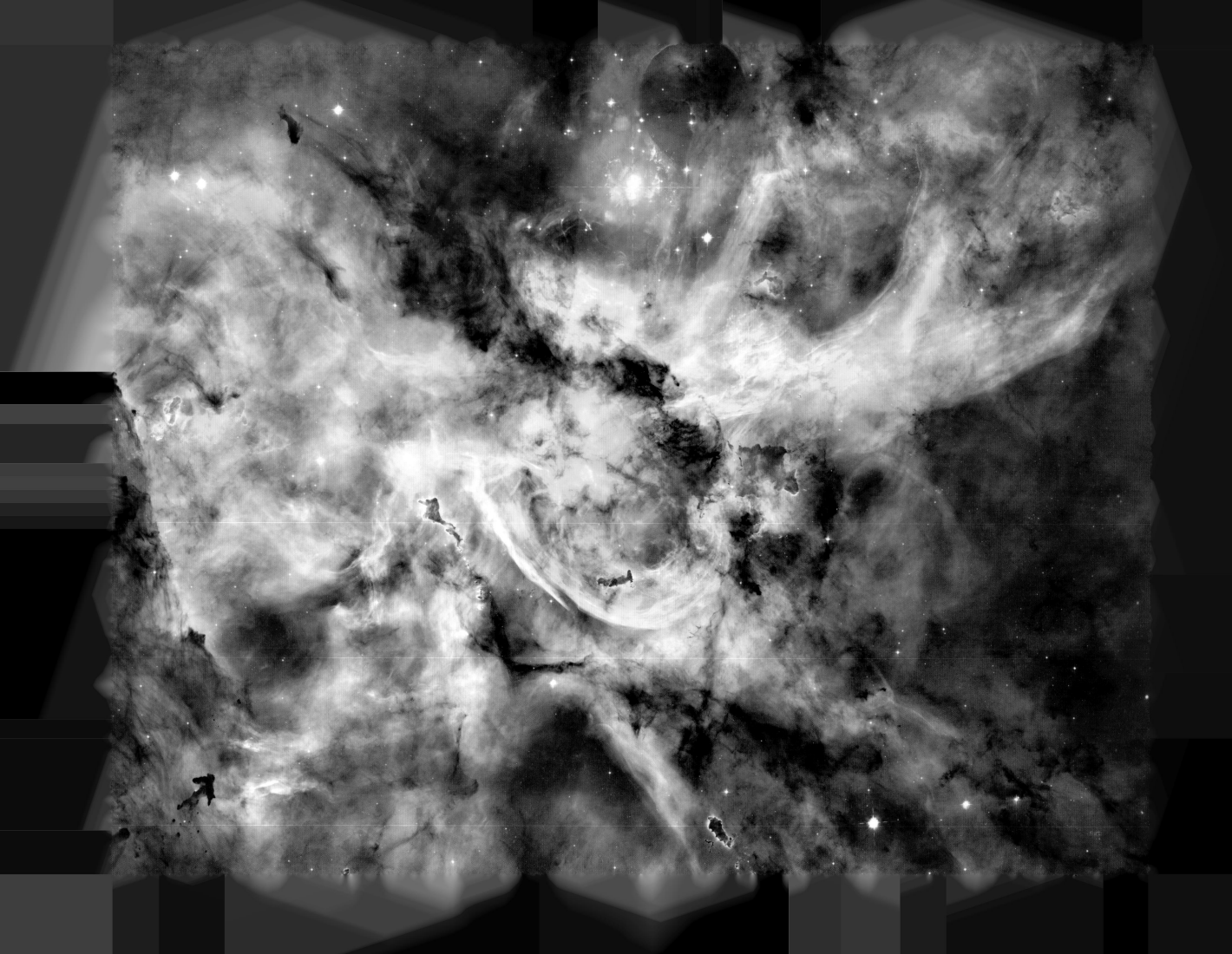

CARINA NEBULA

One of the most spectacular starbirth complexes in the Milky Way, the Carina Nebula lies some 7,500 light years from our solar system, and is one of the jewels of Earth's southern-hemisphere skies. Hubble's space-based views are even better, and the telescope has revealed countless hidden details of its structure over more than a decade of observations.

Spanning roughly 50 light years of space, the nebula surrounds several large clusters of stars that have been born from its material in the last tens of millions of years – relative youngsters in cosmic terms. Among these are some of the largest stars known – supergiants including Eta Carinae, which is surrounded by its own smaller nebula of ejected materials.

Fierce but invisible ultraviolet radiation from these brilliant but short-lived blue stars illuminates the surrounding nebula and causes it to glow in visible light, revealing the delicate forms of multiple veils of gas. A large cloud of cooler gas and dust called the Keyhole Nebula creates a dark shadow across the heart of the nebula where it is silhouetted against the gas behind. Elsewhere, finger-like pillars of dense, dark nebulosity mark sites where new generations of stars are being born today.

CARINA PANORAMA NGC 3372

To celebrate Hubble's 17th anniversary, astronomers stitched together this enormous cosmic landscape encompassing almost the entire Carina Nebula. Hubble's Advanced Camera for Surveys (ACS) instrument was used to photograph the glow of the nebula's hydrogen gas over many separate exposures, while colour information came from images taken at the Cerro Tololo Inter-American Observatory (CTIO) in Chile.

Picture NASA/ESA/N. Smith (University of California, Berkeley)/The Hubble Heritage Team (STScI/AURA)/N. Smith (University of California, Berkeley)/NOAO/AURA/NSF.

EAGLE NEBULA

EAGLE NEBULA

The stunning Eagle Nebula in the constellation Ophiuchus takes its name from the bird-like shape created by shadowy star-forming columns of gas and dust at its heart. It surrounds the bright star cluster Messier 16, about 7,000 light years from Earth, and since it was first photographed by Hubble in 1994, it has become known as one of the best areas for studying the complex processes of star formation.

Fierce radiation from the first generation of stars in M16 illuminates the nebula's cavernlike interior through reflection and emission, while eroding the so-called 'Pillars of Creation' – finger-like tendrils up to ten light years long in which new stars are forming even now. Infrared images of the same region reveal the heat of infant protostars still embedded within the dense pillars, but the nebula may already be past its prime. What's more, astronomers have also discovered that the pillars are threatened by the expanding supernova shockwave from a short-lived star in M16. In another millennium, they believe, this shockwave will tear through the stellar nursery of the pillars, cruelly exposing the newborn stars within, while possibly triggering new waves of starbirth elsewhere in the nebula.

THE EAGLE'S 'SPIRE' M16

A beautiful and delicate star-forming pillar lies in a different part of the Eagle Nebula from the original 'Pillars of Creation', and was photographed by Hubble's ACS (Advanced Camera for Surveys) to celebrate the 15th anniversary of Hubble's launch. Nicknamed the 'Spire', the complex structures within this cloud of gas and dust indicate many different processes at work – dense knots indicate star systems in the act of formation, while streams of escaping gas and a ghostly glow show where the Spire is being carved and eroded by radiation from other nearby stars.

Picture: NASA/ESA/Hubble Heritage Team (STScI/AURA).

ORION NEBULA

ORION NEBULA

Probably the most famous nebula in the entire sky, the Orion Nebula or M42 is also the brightest, clearly visible to the naked eye as a slightly fuzzy 'star' in the sword of the mighty celestial hunter. Even the simplest optical instrument reveals its true nature – a blossoming flower-like cloud of gas and dust illuminated from within by the intense radiation of a small star cluster at its heart.

The nebula lies around 1,500 light years from Earth and has a diameter of about 30 light years, but it is just the most obvious region within the much wider 'Orion Molecular Complex', a huge cloud of star-forming hydrogen that stretches across the constellation and beyond, and bursts into life in several other nebulae.

M42 has proved a rich subject for study by Hubble – the telescope has not only revealed beautiful structures within the nebula in unprecedented detail, but has also studied the stars of the central 'Trapezium', photographed protoplanetary discs around newborn stars, and unearthed a hidden population of faint brown dwarfs that are too small and feeble to shine in visible light.

PANORAMIC VIEW M42 AND M43

This stunning wide-angle view of the Orion Nebula and its close companion M43 (at upper left) combines 520 images taken with Hubble's Advanced Camera for Surveys (ACS) at five separate wavelengths. More than 3,000 individual stars have been identified within the image, ranging from the brilliant monsters of the central Trapezium (each around 100,000 times more luminous than the Sun), to feeble brown dwarfs too small to shine in the same way as the Sun.

Picture: NASA/ESA/M. Robberto (Space Telescope Science Institute/ESA)/Hubble Space Telescope Orion Treasury Project Team.

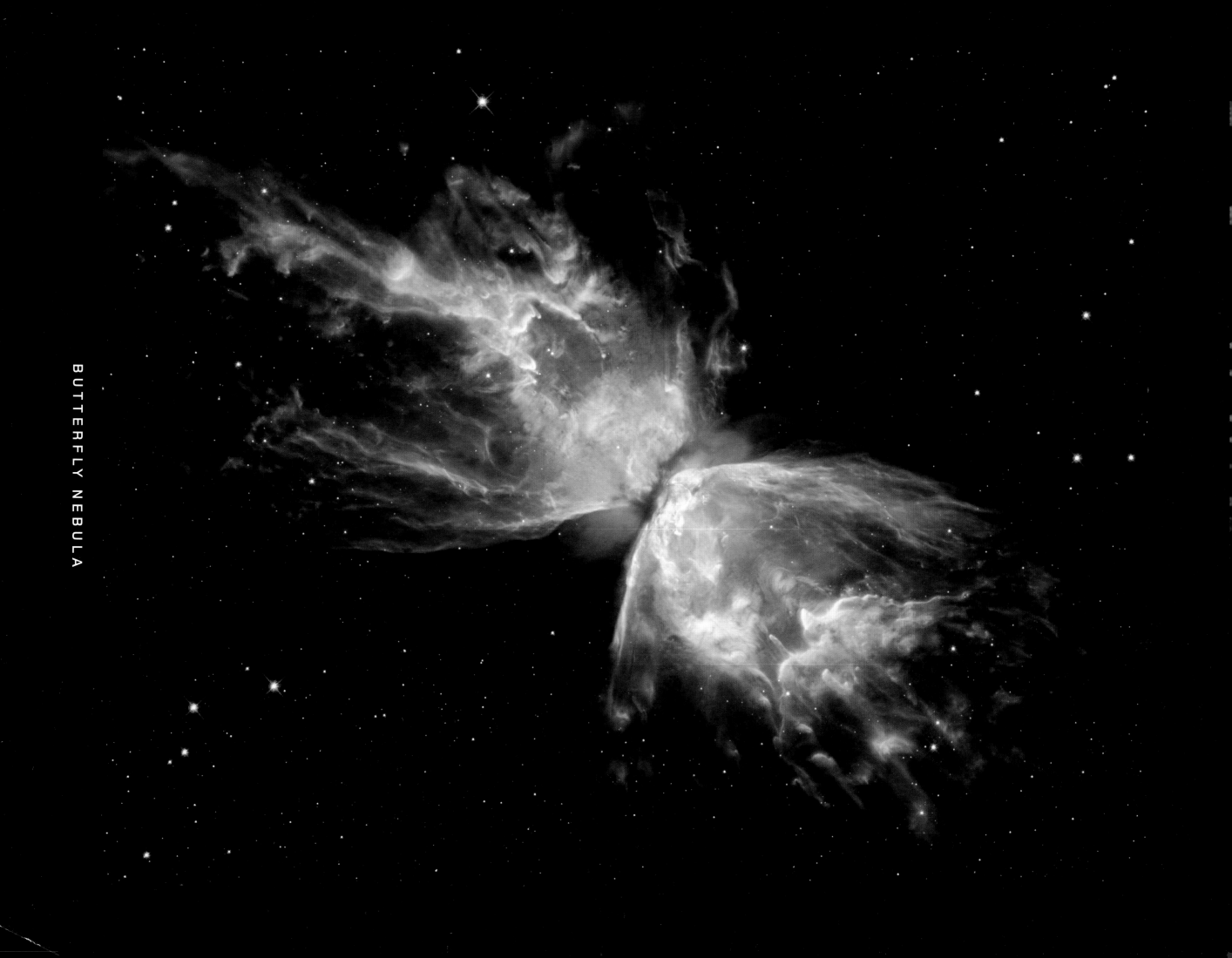

BUTTERFLY NEBULA

BUTTERFLY NEBULA

The beautiful Butterfly Nebula in Scorpius (NGC 6302, sometimes also called the Bug Nebula) is a spectacular hourglass-shaped planetary nebula whose central star, with a surface temperature of 360,000 °F (200,000 °C), is one of the hottest objects in the galaxy.

Material blowing out from the star is pinched around the equator by a dense ring of dust, resulting in two lobes of hot gas that move outwards at speeds of several hundred miles per second. While the current lobes began to form around 2,000 years ago, the central region is surrounded by fainter lobes from a previous eruption.

The Hubble Space Telescope has imaged this nebula twice, revealing more detail than has ever been seen from the ground, but the central star remains obstinately hidden behind the dark dust ring, and its properties can only be inferred from its effect on the surrounding nebula. The dust ring itself, meanwhile, is fascinating in its own right – infrared observations suggest it is packed with chemicals including water ice, calcite, and even carbonate minerals. Since this dust ring is also thought to have formed from material expelled from the central star, it suggests that the chemistry of ageing stellar atmospheres may be more complex than previously imagined.

A NEW VIEW NGC 6302

Following Hubble's 2009 refit, the Butterfly Nebula was an early target for the new Wide Field Camera 3. The result was this impressive view of the nebula's two major lobes, each composed of gas that is streaming away from the central star at speeds of a million kilometres an hour (600,000 mph). Gas within the lobes is heated to around 20,000 °C (36,000 °F) – significantly hotter than most planetary nebulae, and shines substantially in the ultraviolet as well as visible light. This false-colour image highlights glowing areas rich in different elements – the nebula's reddish fringes indicate the presence of nitrogen, while the white areas show where shockwaves are moving through clouds of sulphur.

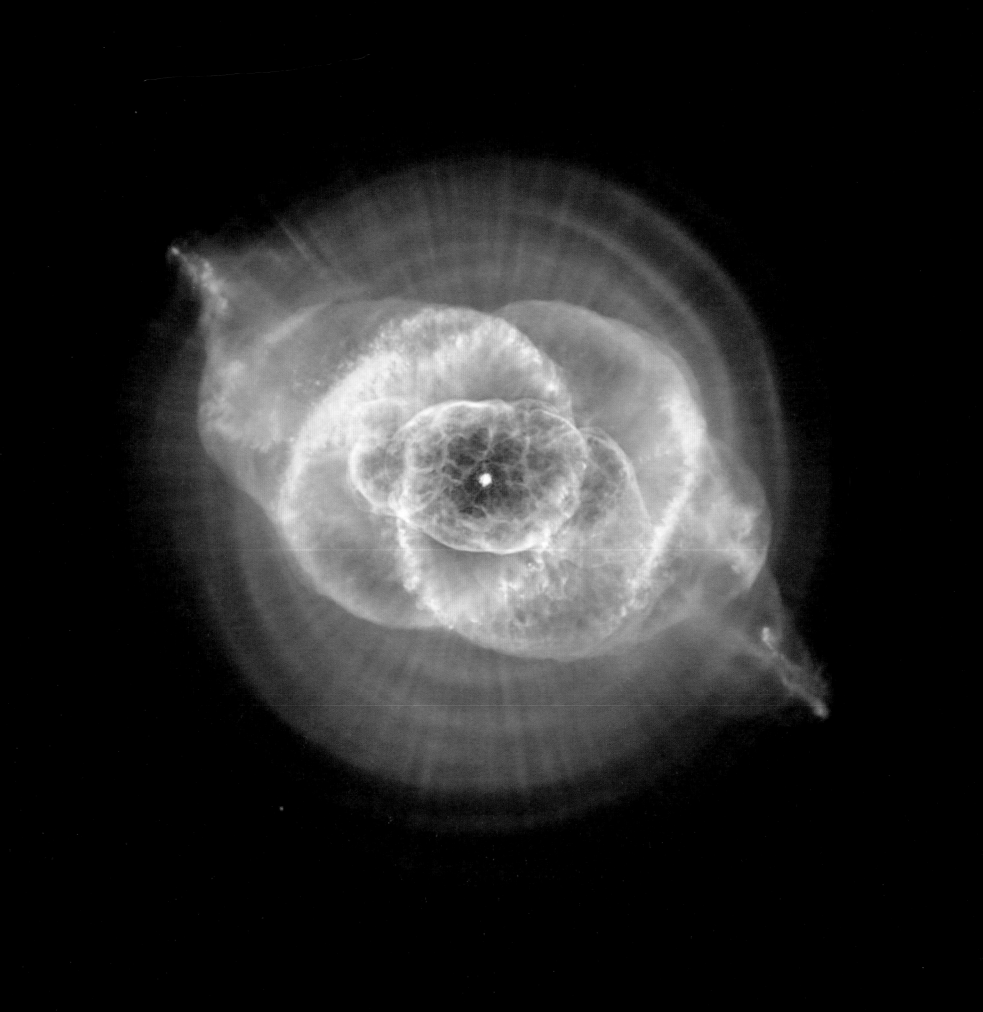

CAT'S EYE NEBULA

CAT'S EYE NEBULA

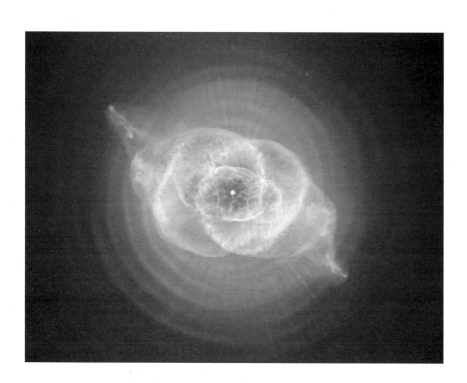

The Cat's Eye is perhaps the most intricately detailed of all known planetary nebulae, a complex mix of expanding concentric shells and corkscrewing spiral bubbles, powered by fierce radiation from its interior and heated by interaction with its surroundings.

Lying around 3,300 light years away in the constellation of Draco, its bright core is about 0.2 light years across and is thought to have formed within the past thousand years. Much fainter concentric rings around the 'eye' must have puffed off the central star in rhythmic pulses several thousand years before, and ragged outlying structures only visible in infrared images may be the result of eruptions around 50,000 years ago.

The most fascinating structure, however, is the corkscrewing series of overlapping bubbles at the centre. Hubble's sibling Chandra has shown the area within the bubbles to be filled with hot, X-ray emitting gas, and astronomers believe the X-rays and the bubbles themselves are a result of a hot 'stellar wind' blowing off the star itself. While the direction and intensity of this wind clearly changes with time, the cause of these changes is still a mystery. However, most astronomers suspect the central star is part of a complex binary system.

CORE OF THE CAT'S EYE NGC 6543

This beautiful Hubble image from the Advanced Camera for Surveys (ACS) shows unprecedented detail of the central Cat's Eye, and reveals a spectacular 'iris' of rings and rays created as the central star threw off its outer layers at roughly 1,500 year intervals. Each ring is in fact the edge of a spherical bubble. The more complex inner shells are created by fast stellar winds blowing out from the star and inflating the bubbles from within – the direction of these winds seems to 'wobble' over time, perhaps due to the pull of a binary companion acting on the central star.

Picture: NASA/ESA/HEIC/Hubble Heritage Team (STScI/AURA).

CRAB NEBULA

CRAB NEBULA

In the 18th century, astronomers discovered a small cloud of light above one of the 'horns' of the constellation Taurus, the Bull. Frenchman Charles Messier made this object number one in his famous catalogue of nebulae and star clusters, and the Earl of Rosse, using the giant 'Leviathan' telescope he built at Birr Castle in Ireland, declared that it looked just like a crab.

Photographs of the Crab Nebula in the early 20th century revealed a remarkable secret – it was expanding rapidly, and appeared to have begun life in an explosion about 900 years before. When researchers looked back through old records, they found that, in 1054 AD, Chinese, Japanese and Persian astronomers had all reported a brilliant new star in Earth's skies. Bright enough to be seen in daylight for 23 days, it did not fade from naked-eye visibility for almost two years.

Today we know that the 'new star' was in fact a Type II supernova – the death throes of a massive star that could no longer support its own weight. The Crab Nebula is the remnant of this enormous explosion, still expanding rapidly, and glowing at temperatures hotter than the Sun. Observations from the Hubble Space Telescope have revealed stunning details of its internal structure and even allowed astronomers to peer into its heart.

The Crab's still-beating heart is a pulsar, a rapidly rotating, superdense 'neutron star' formed by the progenitor star's collapsing core. Powerful jets of radiation shoot from its poles, while shockwaves ripple through the surrounding gas, providing it with enough energy to emit X-rays.

GLORIOUS WRECKAGE M1

Astronomers stitched together 24 separate WFPC-2 (Wide Field and Planetary Camera 2) images to create this spectacular mosaic view of the entire Crab Nebula. False colours are used to indicate the presence of different elements scattered throughout the expanding cloud of stellar debris – blue and red represent neutral and electrically charged oxygen, while green indicates the presence of sulphur.

Picture: NASA/ESA/J. Hester and A. Loll (Arizona State University).

THE MILKY WAY

The centre of the Milky Way, lying in the direction of the constellation Sagittarius, is a source of endless fascination for astronomers. Dense star clouds hide this region from view in visible light, but space-based telescopes (including Hubble) can look through the stars to observe the core in other radiations.

They reveal an extraordinary and violent region of activity embedded in the centre of our galaxy. The area is peppered with heavyweight stellar remnants that pump out high-energy X-rays as they interact with their surroundings. Elsewhere lie huge star clusters containing some of the most massive stars known.

And at the heart of it all lies the mysterious radio source called Sagittarius A*. Astronomers had long suspected that this might be a slumbering 'supermassive' black hole, formed from a collapsed gas cloud early in our galaxy's history, but since every other object in the region maintains a safe distance from 'Sgr A*', it's particularly hard to detect.

However, thanks to observations of how stars around Sgr A* behave, it is now clear that they are right – the core of our galaxy contains an invisible object with the mass of four million suns concentrated in a space smaller than Mercury's orbit. With such a density, this object can only be a truly monstrous black hole.

MULTIWAVELENGTH GALAXY

A spectacular panorama, produced in celebration of the 2009 International Year of Astronomy, combines observations from Chandra X-Ray Observatory (red), Hubble (yellow), and Spitzer Space Telescope (blue) to create our most detailed view yet of the Milky Way's centre. The image reveals a region bathed in X-rays from the central black hole and massive star clusters, an estimated 100,000 stars shining in the infrared, and sculpted columns of gas and dust within which new generations of stars are forming.

Picture: NASA/ESA/SSC/CXC/STScI.

MESSIER 66

MESSIER 66

Picture: NASA, ESA and the Hubble Heritage Team (STScI/AURA)-ESA/Hubble Collaboration.

Galaxies come in a variety of shapes and sizes, with most falling into one of three categories – spirals, ellipticals, and irregulars. However, there are some bizarre and beautiful exceptions. Spirals are probably the most familiar type of galaxy, since our own Milky Way and most of the other large galaxies in our cosmic neighbourhood fall into this category.

Spirals are also the ones with the most complex structures – a central hub of ancient red and yellow stars, a flattened disc of gas, dust, and middleaged stars, and spiral regions with this disc that mark the regions where material is being compressed to give rise to new generations of stars dominated by brilliant but short-lived stars in open clusters. They range in size from about half to double the diameter of the Milky Way, and account for roughly 30 per cent of all known galaxies in our region of the Universe.

In contrast, the other types of galaxy have somewhat simpler structures. Ellipticals are simply enormous balls of old red and yellow stars, resembling the hubs of spiral galaxies. They vary hugely in size, encompassing some of the smallest galaxies and some of the largest, and account for about 15 per cent of all galaxies in our part of space.

Irregular galaxies, meanwhile, are almost structureless stellar clouds, dominated by brilliant but short-lived massive stars, and rich in the gas and dust required to create generation after generation of these giants.

ASYMMETRIC SPIRAL M66

A dazzling ACS view of Messier 66, a spiral galaxy some 35 million light years away in the constellation of Leo, reveals some startling features. The galaxy's spiral arms are asymmetrical and its core appears to be displaced from the true centre of the galaxy. M66 is one of three spirals in a small group called the 'Leo Triplet', and its unusual shape is probably due to the gravitational influence of its two neighbours.

STEPHAN'S QUINTET

STEPHAN'S QUINTET

This beautiful galaxy group, discovered by French astronomer Édouard Stephan in 1877, was the first 'compact group' to be discovered, and has been subject to the most detailed study. Lying in the constellation of Pegasus, it appears to contain five galaxies, but in reality only four belong to a physical group. The fifth (the bluish NGC 7320) is a relatively small foreground spiral galaxy – it lies 40 million light years from Earth, while the others are 250 million light years beyond it.

The four related galaxies consist of the elliptical NGC 7317, the prominently barred spiral NGC 7319, and a pair of interacting galaxies known as NGC 7318a and b. These two show many typical signs of interaction, including loose arcs of gas thrown off by their close encounter, and a burst of star formation where the arcs of gas collide.

Recent studies of the cluster through the infrared Spitzer Space Telescope have revealed more hidden details, including a huge shockwave of hot gas travelling in front of NGC 7318b. For unknown reasons, this galaxy appears to be ploughing through the rest of the group at a speed of around 540 miles (870 km) per second.

A FRESH VISION

Hubble turned the newly fitted Wide Field Camera 3 onto Stephan's Quintet shortly after Servicing Mission 4 in 2009. Using filters to capture blue, green and infrared light as well as specific emissions from hydrogen, astronomers produced a combined image that highlights the colours of different stellar populations.

Picture: NASA/ESA/Hubble SM4 ERO Team.

ANTENNAE GALAXIES

ANTENNAE GALAXIES

Fierce but invisible ultraviolet radiation from these brilliant but short-lived blue stars illuminates the surrounding nebula and causes it to glow in visible light, revealing the delicate forms of multiple veils of gas. A large cloud of cooler gas and dust called the Keyhole Nebula creates a dark shadow across the heart of the nebula where it is silhouetted against the gas behind. Elsewhere, finger-like pillars of dense, dark nebulosity mark sites where new generations of stars are being born today.

One of the most spectacular starbirth complexes in the Milky Way, the Carina Nebula lies some 7,500 light years from our solar system, and is one of the jewels of Earth's southern-hemisphere skies. Hubble's space-based views are even better, and the telescope has revealed countless hidden details of its structure over more than a decade of observations.

Spanning roughly 50 light years of space, the nebula surrounds several large clusters of stars that have been born from its material in the last few tens of millions of years — relative youngsters in cosmic terms. Among these are some of the largest known stars.

ANTENNAE REVISITED NGC 4038/9

In 2006, astronomers uncovered even more detail in the heart of the Antennae using Hubble's ACS (Advanced Camera for Surveys). This enhanced visible-light image highlights star-forming hydrogen in glowing pink. By analysing the properties of enormous 'super star clusters' forming in the collision, researchers discovered that most will eventually disperse, with their surviving stars merging into the wider galaxy, while around 10 per cent will remain intact and develop into globular clusters.

OMEGA CENTAURI

The brightest and largest of more than 150 globular clusters known to orbit the Milky Way, Omega Centauri is so bright that for many centuries it was catalogued as a star. In reality, it is a huge ball of several million old red and yellow stars, 170 light years across and about 16,000 light years from Earth. Like several other globular clusters, Omega Centauri's core is dominated by a giant black hole, weighing as much as 40,000 suns.

Unlike the stellar-mass black holes formed when massive stars go supernova, and the monstrous 'supermassive' holes at the centre of galaxies such as the Milky Way, astronomers still aren't sure how these 'intermediate-mass' holes form.

One theory is that they are created from the merger of several smaller black holes in the densely packed heart of the cluster. More evidence for the crowded conditions here comes in the form of 'blue straggler' stars observed by the Hubble Space Telescope. Astronomers believe that Omega Centauri is around 12 billion years old, so any stars massive and hot enough to appear blue should have long since aged and died. The blue stragglers are therefore a puzzle, but most astronomers believe they form when two lower-mass red stars fall into orbit around one another, collide and merge.

INTO OMEGA NGC 5139

This colourful view was one of the first images made with the new Wide Field Camera-3 following Hubble's 2009 refit. Ultraviolet and visible data have been combined to exaggerate colour differences, revealing a wide range of star types. Most are sedate yellow-white stars, similar to our own sun. Bright red stars are swollen giants near the end of their lives while most of the blue stars are dying stars throwing off their outer layers to expose a hot interior that shines more brightly in the ultraviolet before fading away.

Picture: NASA/ESA/Hubble SM4 ERO Team.

SOMBRERO GALAXY

SOMBRERO GALAXY

Lenticular galaxies are best described as spiral galaxies that have lost their spiral arms. Viewed edge-on they appear to be lens-shaped collections of stars, hence the term lenticular galaxy.

The Sombrero Galaxy is the dominant member of the M104 group, shepherding at least seven other galaxies. It is officially classified as a spiral galaxy with tightly wound arms; however, it looks extremely similar to many lenticular galaxies. Built up of a staggeringly large number of stars – astronomers estimate 800 billion – most of them are crowded into a central lens-shaped region spanning nearly 50,000 light years.

BRILLIANT CORE M104

This magnificent Hubble image reveals a spectacular profile of the Sombrero Galaxy. With a mass equivalent to approximately 800 billion Suns, this galaxy gets its name from its unusually brilliant and bulbous core and the dark gas lanes seen edge-on in its disc. X-ray emissions suggest that at its heart the Sombrero Galaxy harbours a black hole with the mass of a billion Suns.